YOUR AMAZING BRAIN

DWAYNE HICKS

New York

Published in 2023 by The Rosen Publishing Group, Inc.
29 East 21st Street, New York, NY 10010

First Edition

Editor: Greg Roza
Designer: Michael Flynn

Photo Credits: Cover, pp.1 Anurak Pongpatimet/Shutterstock.com; p. 5 MillaF/Shutterstock.com; p. 7 metamorworks/Shutterstock.com; p. 9 sciencepics/Shutterstock.com; p. 11 rbkomar/Shutterstock.com; p. 13 Triff/Shutterstock.com; p. 15 peakanucha/Shutterstock.com; pp. 17, 19 CLIPAREA I Custom media/Shutterstock.com; p. 21 Monkey Business Images/Shutterstock.com.

Cataloging-in-Publication Data

Names: Hicks, Dwayne.
Title: Your amazing brain / Dwayne Hicks.
Description: New York : Powerkids Press, 2023. | Series: Your amazing body | Includes glossary and index.
Identifiers: ISBN 9781725339576 (pbk.) | ISBN 9781725339590 (library bound) | ISBN 9781725339583 (6pack) | ISBN 9781725339606 (ebook)
Subjects: LCSH: Brain–Juvenile literature. | Nervous system–Juvenile literature.
Classification: LCC QP376.H54 2023 | DDC 612.8'2–dc23

Manufactured in the United States of America

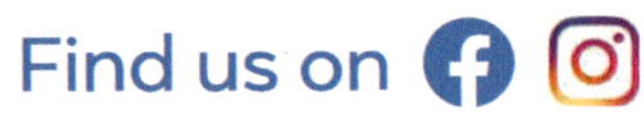

CONTENTS

Use Your Brain!

Your brain is truly amazing. It controls everything you think, say, and do. It allows you to run and jump, but it also allows you to sing and draw! It forms memories and **emotions**. Your brain is more powerful than any man-made computer!

The Nervous System

Your brain is part of your nervous **system**. Your body has many nerves, which send messages to the brain. Your brain is connected to the **spinal cord**. This is the main pathway for information traveling between body parts and the brain.

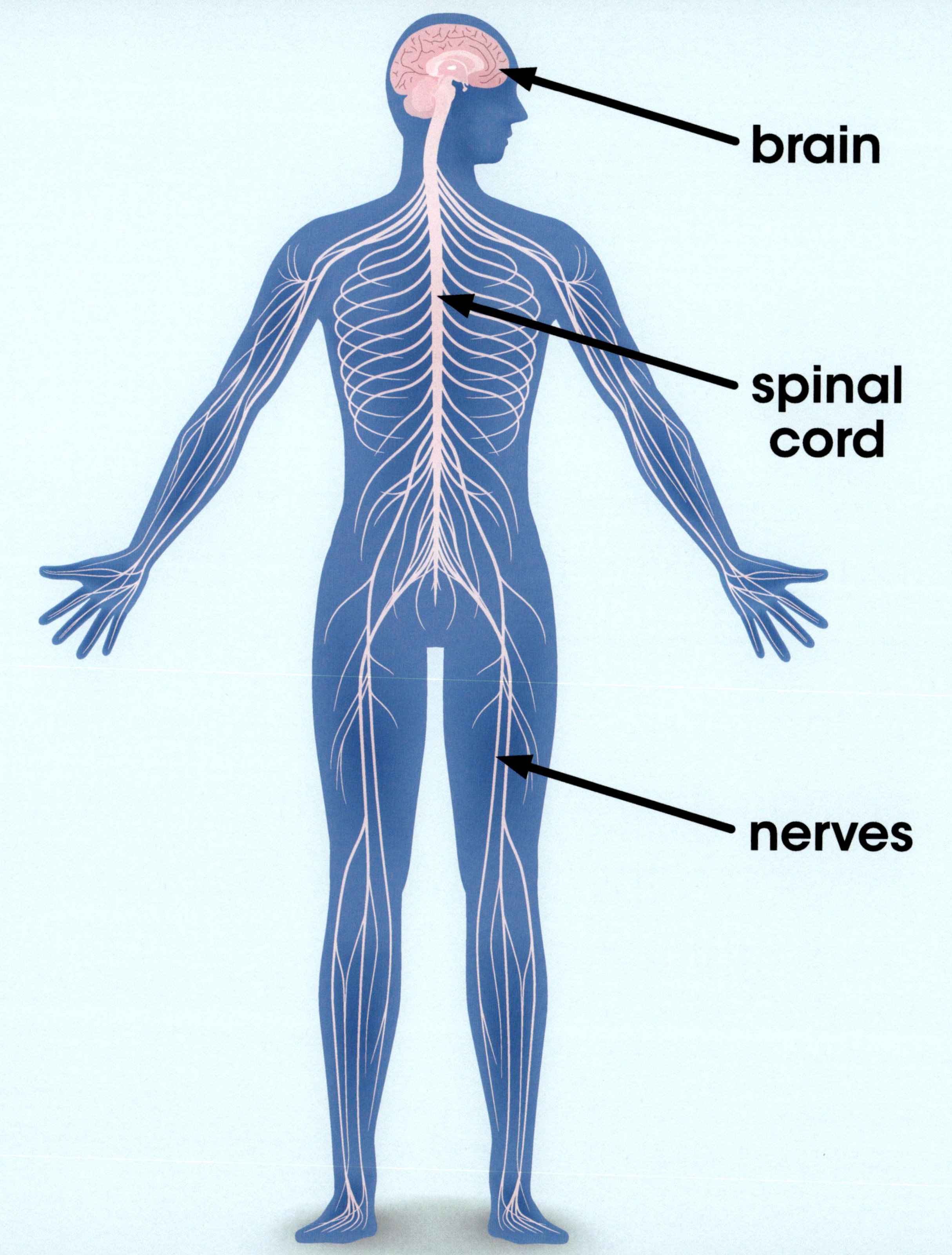
brain
spinal cord
nerves

It's Electric!

Your senses—sight, smell, hearing, taste, and touch—gather information from the world around you. This information travels through nerves to the brain in the form of electrical **signals**. This is similar to the electricity that powers computers and lightbulbs.

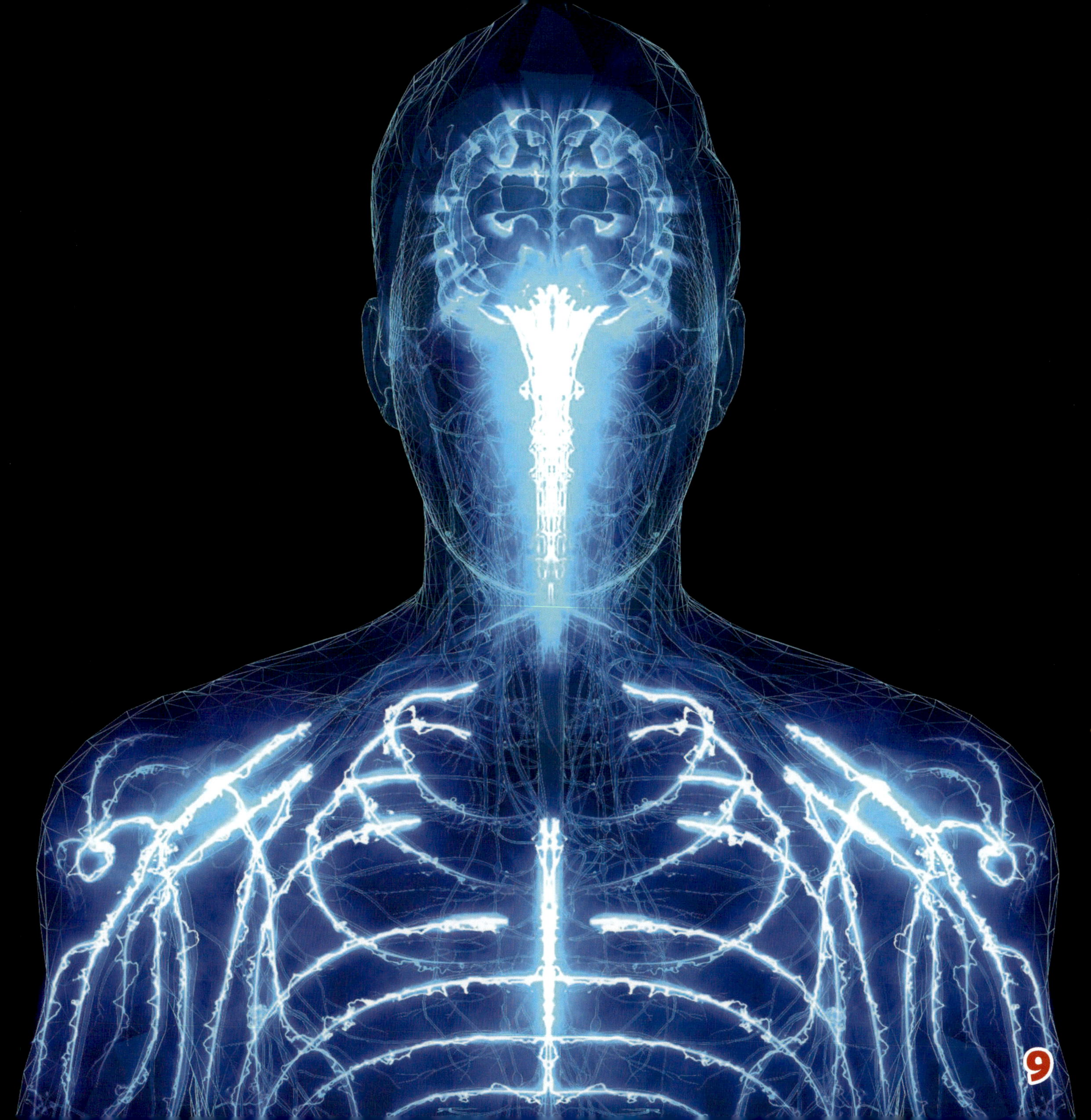

Gathering Information

The brain "reads" the electrical signals our senses gather. For example, it tells us when something is too hot to touch. In response, the brain sends a signal back to the hand in the form of pain. This pain tells you to move your hand—quickly!

Two Halves

Your brain has two halves. The right side controls the left part of your body and performs creative activities, such as art and music. The left side controls the right side of your body and performs **logical** activities, such as math and science.

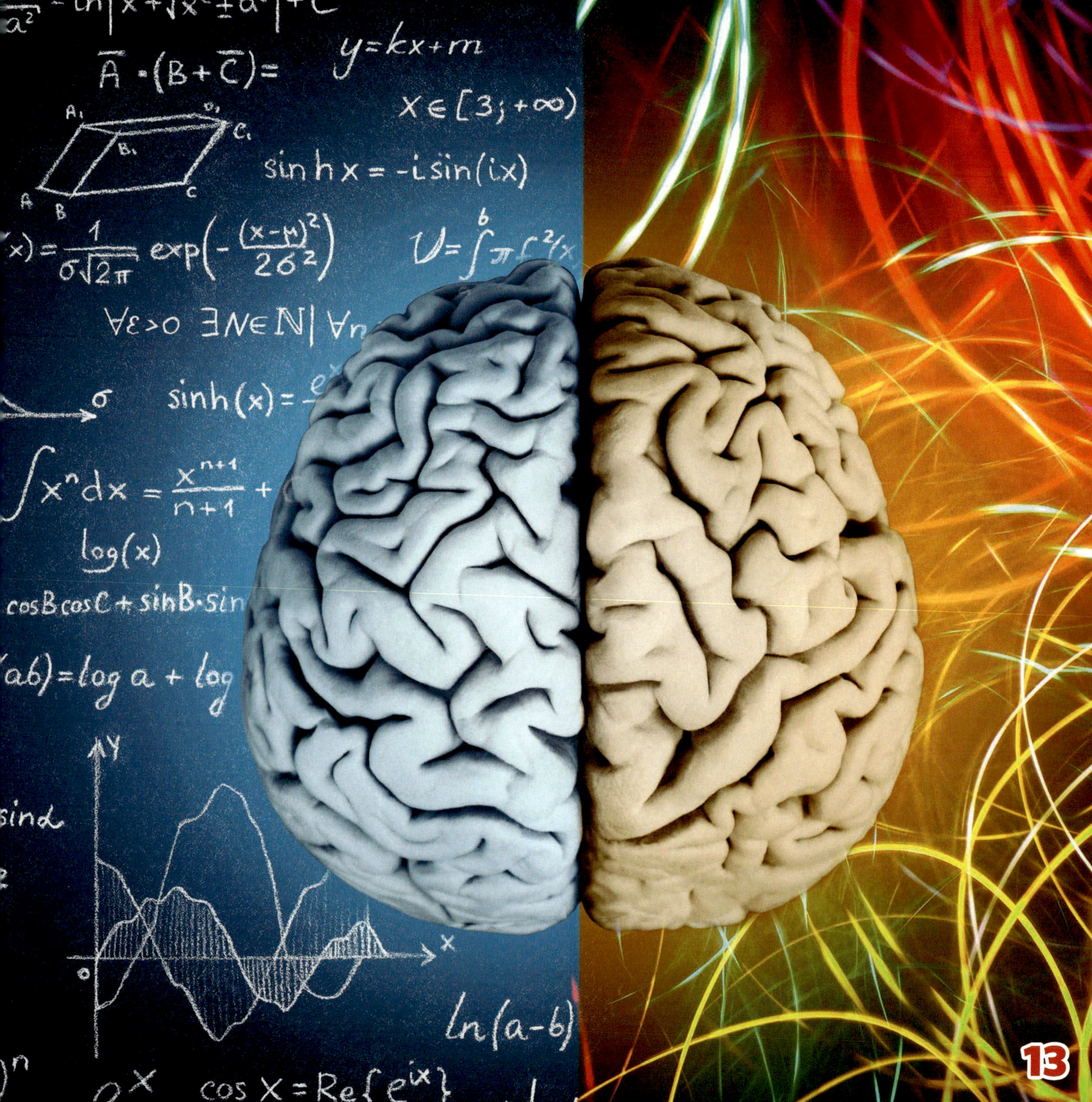
y=kx+m
X∈[3;+∞)
sinh x = -i sin(ix)
log(x)
ln(a-b)
cos X = Re{e^{ix}}

Different Areas, Different Jobs

Different areas of the brain have different jobs. One area of the brain helps form memories and helps us learn. Another area is in charge of emotions. Another area deals with **motor skills**, which include learned activities like riding a bike.

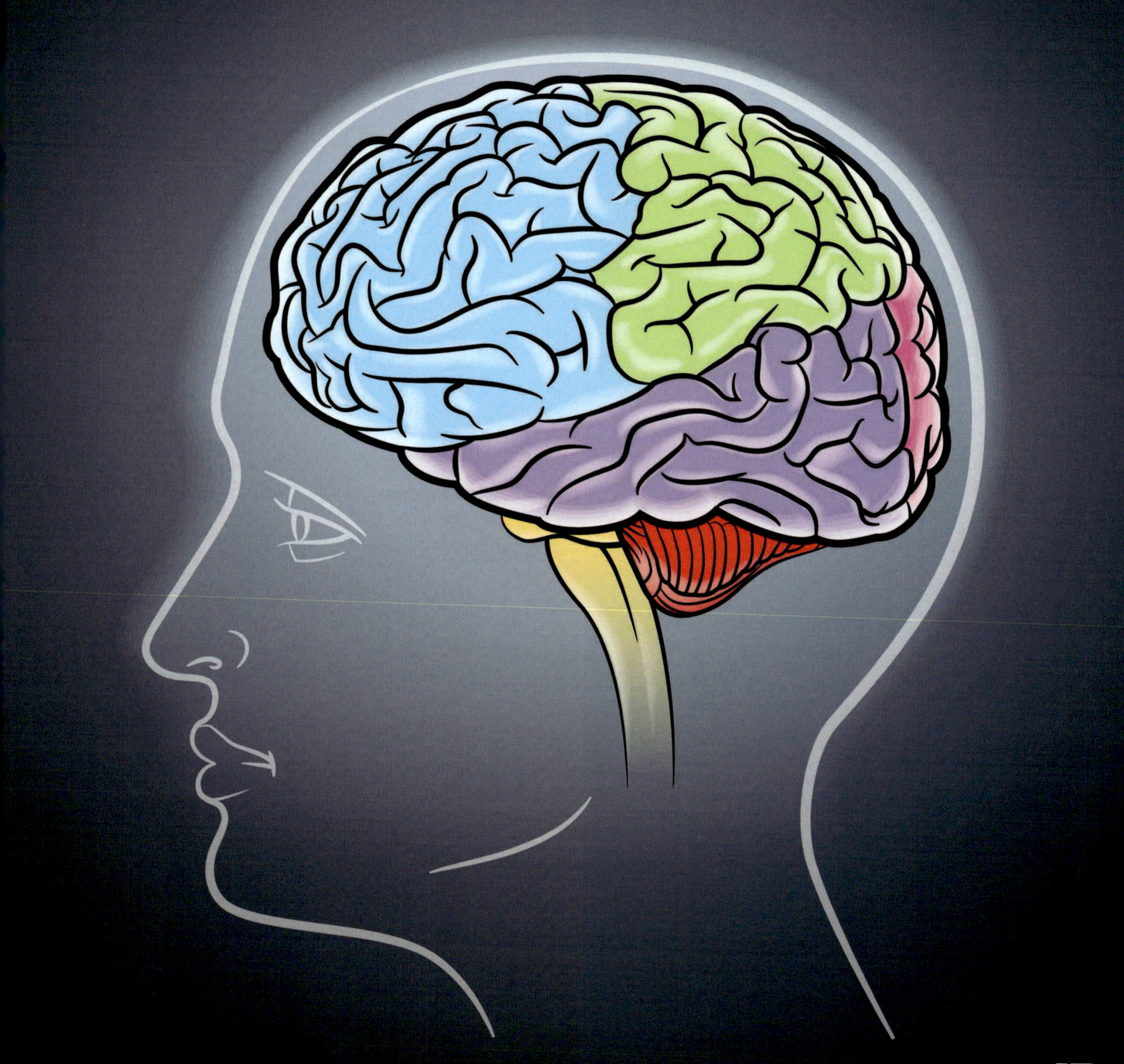

The Brain Stem

The brain stem is the part that connects your brain to the spinal cord. Your brain stem controls body functions that you may not even be aware of! Your brain stem keeps you breathing. It keeps your heart pumping too.

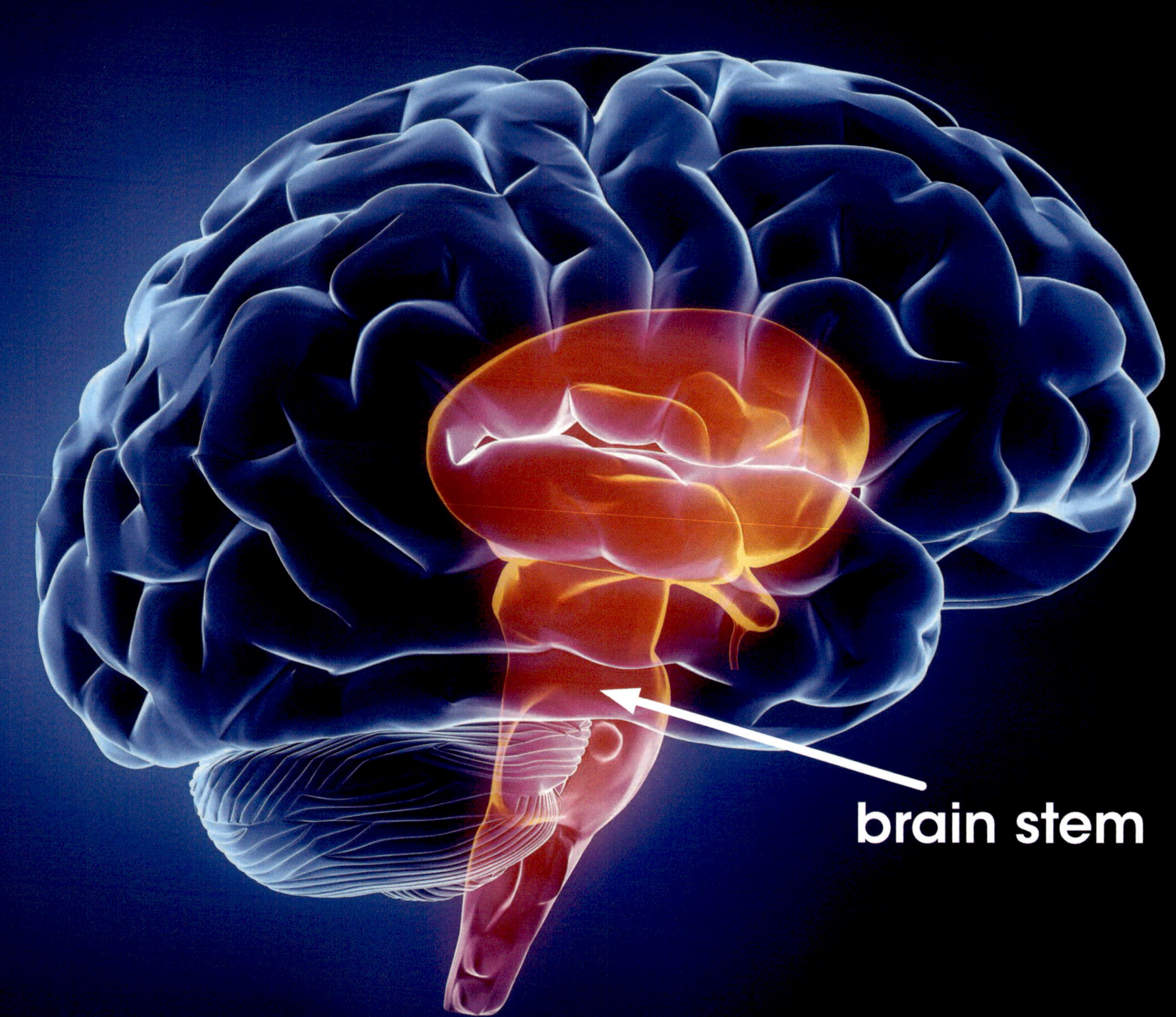
brain stem

At the Center

Inside your brain near the brain stem is a part that controls how hot or cold your body is. The pituitary **gland** is also near the brain stem. It helps control human growth among other things.

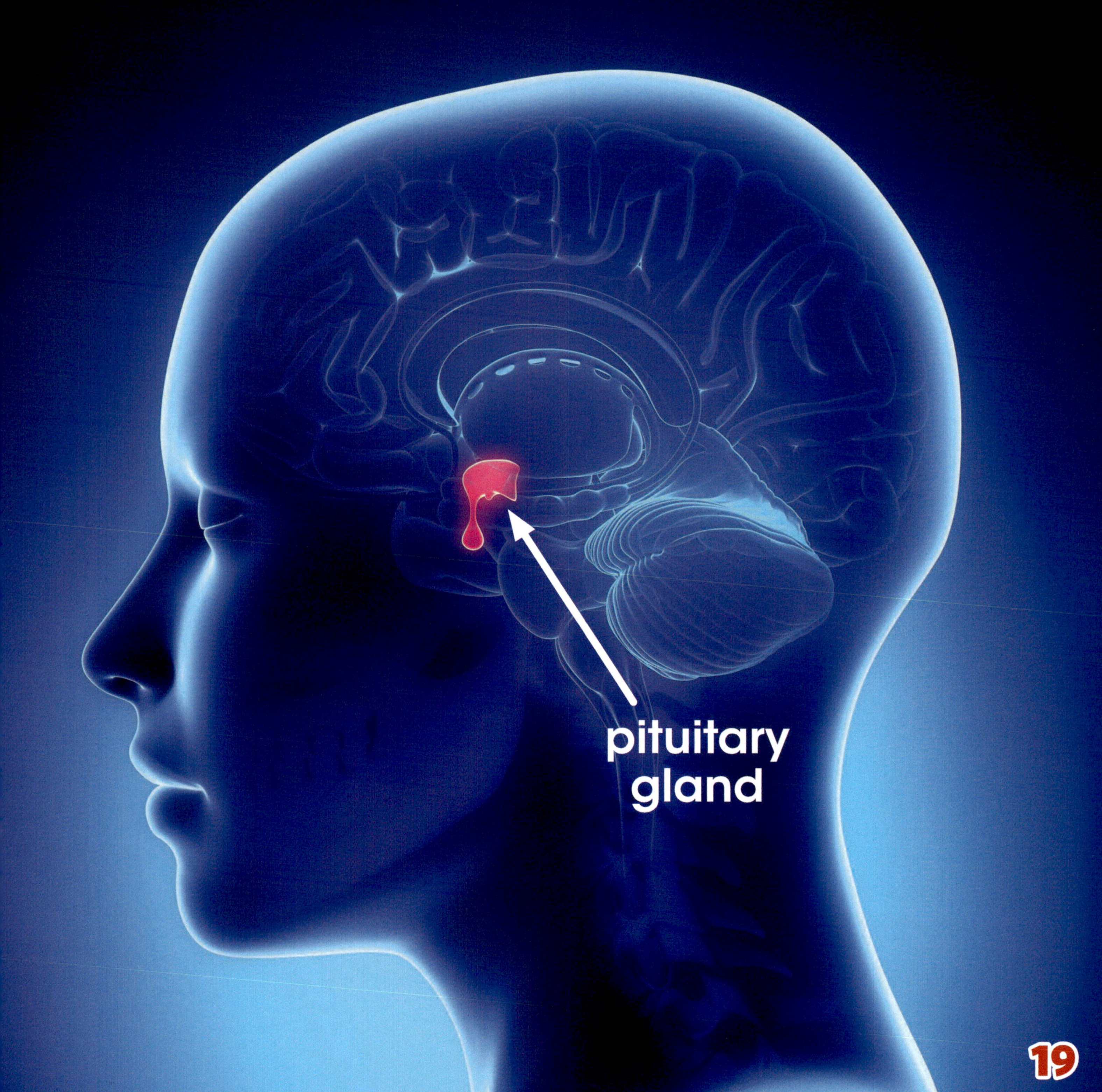
pituitary
gland

Your Healthy Brain

It's important to keep your brain healthy. Don't do drugs or drink alcohol. These things can have long-lasting effects on the brain. It's important to eat a healthy diet. Don't forget your helmet when you ride your bike!

GLOSSARY

emotion: A feeling, such as joy or anger.

gland: A part of the body that produces a substance to be used by the body.

logical: Having to do with the processes used in thinking and reasoning.

motor skill: The body's ability to manage the process of movement using muscles.

signal: A message; something that results in an action.

spinal cord: The large bundle of nerves that runs through the center of the spine.

system: A group of body parts that work together to perform an important function in the body.

FOR FURTHER INFORMATION

WEBSITES

Biology for Kids: The Human Brain
www.ducksters.com/science/brain.php
Read more about the brain, see a helpful diagram, and find links to other realted topics.

Your Brain & Nervous System
kidshealth.org/en/kids/brain.html
You can learn even more about your nervous system at this detailed website.

BOOKS

Christensen, Evelyn B., and Susan E, Christensen. *The Amazing Brain Book for Kids: Brain Games, Logic Puzzles, Riddles & More!* Emeryville, CA: Rockridge Press, 2021.

Seluk, Nick. *The Brain Is Kind of a Big Deal.* New York, NY: Orchard Books, 2019.

Publisher's note to parents and teachers: Our editors have reviewed the websites listed here to make sure they're suitable for students. However, websites may change frequently. Please note that students should always be supervised when they access the internet.

INDEX